IGNORANCE IS ILLNESS

IGNORANCE IS ILLNESS

Disproof Of The Big Bang Theory Through Healing

By Eric A. Staubes
The QcCubed Space Coach

To order additional copies of this book, contact:
Xlibris Corporation
1-888-795-4274
www.Xlibris.com
Orders@Xlibris.com
128439

Contents

Welcome, Fellow Traveler, aboard the Starship Earth. The tour guide you are reading has been prepared with one particular goal in mind. That is to get the traveler from Point A to Point B in the most cost effective and elegant manner possible. Point A is our current collective understanding of how this Universe works and Point B is where we know how it really works. For those of you who know you are interested in learning more, this is an exciting journey. We can already predict (because we know from experience) that when your relationship with space epiphanizes, you will know the importance of this learning and you will be glad. If you're not sure whether you should be interested or not, just sit back, relax and enjoy the ride. Excitement is like laughter; it is contagious. Your Mother Earth is changing and she wants to get your attention; which she will do, through what is perceived as one disaster after another. Soon you, too, will agree to become interested in the destiny of yourself and your loved ones on this Starship Planet Earth, it will be your turn to grasp the meaning of space, and you, too, will be glad.

My job as the QcCubed Space Coach is to teach my team how to win. As you need my help just ask. We shall overcome any and all limitations. We are the pride of the Universe.

Thanks . . . . Coach Staubes

First Chapter

ONE

I am the Space Coach and I invite you to join me on a voyage through the Universe. By definition there can only be one Universe; coming from the Latin unus—"one"—and versus, pp. of vertere—"to turn"; literally, "turned into one." The vehicle we are using to move through space is our beautiful imagination. Einstein said, "Imagination is everything. It is the preview of life's coming attractions." So come along with me on a happy voyage to the planets and then off to the distant stars, traveling through the space that contains them; or on a journey into the nucleus of the smallest atoms, through the space that contains the particles; realizing these are different portions of the same one space. Macro space; micro-space; inner space; outer space; different terms; different places; same space.

Before anybody can make any valid claims about the Universe, they have to at least know what space is and how it works, since everything in the Universe is contained in one vast space. The space that contains us is the same as the space that is contained within us. Because there is only one space, it contains everything. As we uncover the secrets about space, we will all marvel at the power the Universe makes available to us. Right now we are being handed a story about the Big Bang that supposedly started this Universe many billions of years ago and the fact is, that's rubbish!

The Universe is infinite. That means it has no beginning and it has no ending, either. In a separate writing I have shown how the system of thinking we learn (because we are all born into it), is based upon the validity of human perception, which sees three-dimensional objects floating through empty space and builds an entire philosophy of life based upon those observations. The problem is, matter is not solid and space is not empty. But every time we fall down, we are reminded of just how solid matter is. And when we fell, whether it was from a horse or from a lawn chair, we fell through empty space until we hit something solid.

Obviously I'm not just talking about the way things <u>seem to be</u>. No, I'm talking about the way things <u>really are</u>. All the particles that make up matter; the electrons, protons, neutrons, quarks, leptons, gluons, bosons; the list seems endless; nonetheless, these particles are all moving around at tremendous speeds, through the one space, below the level of our perception; building up all the objects we call material reality. Where we perceive solid matter, there is really motion. Where we perceive empty space there is energy. Space is full of energy fields which we cannot see. We can only see what the energy fields do. They move particles. These particles form solid matter, but they never stop moving. That is to say, they never stop being moved by the energy fields in space. In other words, space moves matter.

Come along now and board the spaceship I've built for us with the one imagination that permeates all space. You have your own imagination and I have my own imagination because you are you and I am me. But let me ask you something. If there is only one space (and as we travel out into the planets I hope to convince you that there is only one space) then it must be that the space that contains you and your imagination is the same space that contains me and my imagination. How are you and I connected? We both live on the same planet and we breathe the same air; sharing the one space that fills the Universe. Our minds are linked by the space that contains us.

A convenient way to think about the energy of space is in terms of zero point fluctuation. Let me explain my interpretation of what that is. Because space looks empty we thought it was empty. Hence, out in deep space where there should be no energy (because there are no planets or stars in much of deep space; hence, no physical source of energy fields, like gravity); out where there should be no energy there is still energy! Four degrees Kelvin, to be exact. Where there should be no heat, there is still fluctuation of energy. Random fluctuation, mind you; which means it accomplishes nothing. But why is it there? Where does it come from? What does it do?

The Big Bangers claim this zero point fluctuation is just energy left over from the Big Bang. Once again, hogwash! This Universe has been around forever. There's no big bang to be left over from! So where does this energy come from? The simplest answer is, of course, that it comes from space itself. Others have suggested this and I am here to prove that the fabric of space contains the energy that moves the particles through space; not unlike how, on the macro level, the energy fields move the planets through space. Why is this so important? Why should you, or anybody else for that matter, care where energy comes from? What's the difference if it does or does not come from the Big Bang? I mean, after all, these are the scientists that put men on the Moon and robots on Mars. Haven't they proven that what they say must be true?

Since modern day physicists are responsible for all the wonderful technological advances we enjoy, including NASA and our flights into space, if they believe the Universe began with a big bang, haven't they proven this with all the progress they've made? And the answer is, <u>no</u>. And that's an emphatic <u>no</u>! A very important <u>no</u>. Here's why it is important. If you believe in the Big Bang you cannot believe in an infinite Universe. And as long as you believe in a finite Universe, you lack the capacity to cohere the random background fluctuation of the energy of space. This lack of control is what subsequently allows for the random occurrence of energy deficiencies in the human body and it is these deficiencies which cause all degenerative illness. Another way to hand each individual the blame for their own degenerative illness is to assert that in an infinite Universe there is always energy available for us to use to craft our own well being; whereas, if there is a beginning there was a Big Bang and that's all the energy there is! It's all been distributed. You are just another victim!

We only think there has to be a beginning because that is what we were taught. Now, we have to trust our own senses, otherwise we would just go insane. If we all see the same event, we all expect everyone else to more or less agree with what we saw. If you and I and four others see something happen and five of us agree with what we saw but you say you saw something else, we're all going to have to think about you just a little bit differently. That is because we all learn to trust our own perception and we verify it by the agreement of others. We have to for the sake of our own survival. And the system of thinking that processes and stores and communicates the information about our perception we can call the philosophy of three-dimensionality, because it is founded upon our own perception. We measure objects in length, width and height and it is pretty obvious that all objects begin here and end there. The table we sit at, for instance, begins on this side of the table and it ends over there, on the other side of the table. The car we drive begins at the bumper under the hood and ends at the bumper under the trunk. For sure, everything has a beginning and an ending. That is, everything in a three-dimensionally described, finite Universe. But there is more to the Universe than meets the eye!

Our job, as citizens of the Planet Earth, is to figure out what our relationship with the infinite Universe really is. This treatise represents an interpretation of our perception which is slightly different from what we have all learned growing up. Life is not the way it appears to be, it is the way we <u>think</u> it appears to be. That subtle difference means on the one hand, we sit back and watch what happens to our planet as the climate changes globally, or on the other hand, we actively participate in the direction our planet evolves in. We have more responsibility for our destiny than most folks realize.

The following is a brief description of the Big Bang Theory. Our reference source throughout this writing is Wikipedia, the free encyclopedia offered on the World Wide Web. When we come to a word or a concept that needs to be defined, we "Google it" and take from the choices offered the Wikipedia define—tion, as follows. "The Big Bang Theory is the prevailing cosmological model that describes the early development of the Universe. According to the Big Bang theory, the Universe was once in an extremely hot and dense state which expanded rapidly. This rapid expansion caused the Universe to cool and resulted in its present continuously expanding state. According to the most recent measurements and observations, the Big Bang occurred approximately 13.75 billion years ago, which is thus considered the age of the Universe."

We at **QcCubed Scientifics** disagree. Our argument against that cosmo—logical model happens also to be the healing protocol that forms the information—based therapy designed by QcCubed to remove the symptoms of all degenerative illness from our global experience here on the planet Earth. That sounds like a lot to do but we're going to do it. We have found the secret of the Universe and we are sharing it with everyone on the planet through the World Wide Web of information that now forms the new consciousness of this Spaceship Earth as it sweeps through space.

The Big Bang implies the beginning of the Universe and that is impossible for something which has always existed. The flaw is in our thinking. In a world defined by solids, everything has dimensions which start here and end there, so there has to be a beginning. But that's a man-made restriction; invented by physicists to accommodate proven theories, like Einstein's General Relativity. What we propose is that there is another way to account for observed phenomena. Since we live in an infinite Universe, it never ends. It has no beginning. It doesn't need one. There is a much easier way to explain the expansion of the Universe.

When I first heard about the Big Bang Theory I was in seventh grade Earth Science class in 1961. Maybe the Universe was thought to be about six or seven billion years old back then. We were told that in the beginning, the Universe was very small. All the matter from all the stars and galaxies was actually compressed down into an infinitely dense object that was only as big a little pea. Imagine that! The scientists that know everything have decided the Universe, as vast as it is, used to be as small as a marble. They came up with that idea to explain why the Universe is expanding. Wow! What an act of genius! Here's their logic. The Universe is constantly getting bigger. It is expanding in all directions. If it's bigger today than it was yesterday (and it is) then that means it was smaller yesterday than it is today; and even smaller last week and the week before that. So here's the brilliant conclusion they came up with. Since it gets smaller and smaller as we go back week by week and year by year, that means if we go back far enough, we get to where it all began. The further back in time we go, the smaller it gets. And if we go back far enough, since the Universe according to these calculations is getting smaller and smaller, eventually it has to go to nothing. Instead of nothing we will call that imagined state a singularity, which means it is infinitely dense.

C'mon man. That's ridiculous. The only thing that is infinitely dense is the mind of the scientist who insists that's how things in this magnificently beautiful Universe actually came to be. In this discourse we provide another explanation. It's a lot easier to fathom and it is easier to prove, because the proof is in the healing. This essay is the text for the complete healing of all degenerative illness from the human experience of the planet Earth.

Second Chapter

CAUSE

Behind every effect there is a cause. We learn this in our early, formative years. For instance, the biggest kid in our group we nicknamed "Odd Job." We used to have fun putting him on one side of the seesaw and two of us regular kids on the other side so we could make the board move. If one of us got off while our side of the teeter-totter was down and Odd Job's side was up, he came down fast, catapulting the other remaining kid up in the air. That was cause and effect. We didn't have a name for the principle involved. We wouldn't learn that for another couple of years. But we had the experience.

It looks like that event, or any event for that matter, can be described by two elements; one being cause, the other, effect. But that leaves out an important third element, namely, the action itself. Experience tells me we need a good, sound understanding of the way a principle works before we go flying off into space. It is only by fully understanding the principles of space travel that we can be confident of success. Behind all effect there is cause and action. Thus, whenever we consider any event, we must think in three's. Since the Universe has always existed, we don't have to be concerned about where it came from or how it started. Instead of looking for its cause, we will spend our time considering it <u>as a cause.</u>

Cause may not be something we can directly perceive. Imagine falling off the bicycle you are riding and scraping your hands and elbows on the pavement. You could say that the cause of the scrape is the physical act of falling off the bike, but if there were no gravity, or even less gravity, like riding a bike on the moon; when you fell, you'd hit the ground so gently there would be no scrape. So the impact, caused by the pull of gravity; that's the cause of the scrape. And you can't see gravity.

We cannot see gravity, we can only see what it does. Is that true of all cause? Perhaps. Is all cause invisible? Of course, we can see effect. We

14

know that. The question is, can we see cause? If we cannot see gravity and gravity is the cause of many of the actions we witness in the physical world, then at least some portion of all cause is beyond human perception. On the other hand, since gravity affects anything with mass, and everything in our physical world has mass, then it can be asserted that all events that transpire within a gravitational field owe their cause to energies which we cannot directly perceive.

But of course, you and I are going to be traveling outside the Earth's gravitational field. Most of the time we are out in space we will be weightless. But there is still a gravitational effect between you and me; then between you and me and every other passenger on the ship; as well as between all the passengers _and_ the ship; because wherever there is matter there is mass and between any and all accumulations of mass, there is gravitational attraction. That means, wherever there is matter, we can see the effect; which is the mass, in whatever form it occupies; but we cannot see the cause, since gravity is invisible.

We cannot see space, either. Now maybe that is just a coincidence. Or maybe it isn't. Maybe it's the single most important thing any one of us could learn in this wonderful experience of life we all share. It seems as though whatever we see or become involved with in the physical world is an effect, which is caused by energies moving through space which we cannot see. Cause is invisible. Effect is visible. Let's take it one step further. Whatever we can see is an effect on the material plane of existence and all the effects are caused by energies we cannot perceive; which are moving through space. All the events that transpire in space are caused by the energy of space. Some people I've spoken with find that concept very threatening; as if they are solely responsible for their own universe and they do not need or want any competition from a unifying principle. Please relax. The Truth about space is very comforting and enjoyable, to say the least. If it were not so I would tell you.

Let's fly by the Moon first. Then to Mars, to see how the Curiosity Rover is doing. Then I'd like to go out to see the Moons of Jupiter. Now these are some fairly long distances we'll be rocketing through space. But you realize, of course, that the space between Earth and Moon is the same space as that between the Moon and Mars; all part of the one space that contains our solar system. "Rocketing through space" is a metaphor. Rockets are what we used to use to travel through space. But they won't do for a trip to Jupiter. They're too slow. We don't want to take years to get there. With the new system of space propulsion it should take less than two weeks.

What we are tasting is the future we are going to create—first in our imagination—in the invisible world of cause; followed naturally by being created in the physical world. So we have cause and effect; one and three.

The Universe works in three's. What is number two? It is the action. Cause—Action—Effect. One—Two—Three. Another way to think of it is as Energy—Motion—Form; or Idea—Thought—Form. Or, "I want a house;" architect draws plans; contractor builds house. And maybe you're thinking, "I want a house" is the cause, since those four words are visible, and they are the cause; so maybe some cause is visible? But the words are not the cause. They represent the desire; the idea formed by the desire. That is the cause; once again, invisible.

Everything we see is created by everything we do not see. We see that which is contained within space, but we do not see space itself. And everything in the Universe is contained within one space. Where there is motion, there is the action set into motion by the cause. The cause, through motion, manifests the effect. One—Two—Three. In this Universe all the particles are moving through space. But where does the energy come from that moves these particles? According to the "Big Bangers," there was an explosion many years ago that started the Universe. But that theory is so easily disproved, it is actually beneath me to do so. It is much more to our advantage to present the correct explanation of where all energy comes from, because it comes from space itself.

Let's not be concerned right away about the mathematical proof of our primary assumption concerning space. The idea of an eternal Universe is much easier to accept than that silly notion about an infinitely dense little piece of matter the size of a pea all of a sudden exploding. This is a good time to consider Occam's Razor, otherwise known and revered in the scientific community as the law of parsimony or succinctness. This is the principle we are encouraged to use when selecting from competing hypotheses: choosing the one which makes the fewest assumptions. Big Bangers want us to choose the finite over the infinite. They don't realize it but they are only demanding a beginning because they think there has to be one based upon the philosophy of three-dimensionality. Without the need for a Big Bang, the infinite Universe is very easy to understand. But there is something even more important than that! When we don't restrict the Universe with our own limited thinking, it invents ways to share more of its identity with us. Remember: the Universe is One . . . the energy of light passing through the seam in space is Two . . . and all the human beings on this planet, you and me and everybody else, we are number Three.

Instead of trying to imagine all these billions of galaxies, each with billions of stars, somehow all owing their origin to a time when there was no Universe (we have no logical right to assume there could be no Universe) and all their mass was reduced to a tiny little pea-sized something of infinite density (how ridiculous!) just imagine the Universe is infinite, which means it has always existed. What is so difficult about that? It's easy. It's the way a kid thinks;

no complications unless added by someone else's ideas. Scientists figure everything out by having questioning minds. That's good. But what is so difficult about a Universe that has always existed? Only this: three-dimensionalists, or materialists, cannot accept something which they cannot control. Rather than just assume there are principles at work in the Universe that they might not understand, they insist upon creating the Universe themselves, first in their theories, then followed up by proof in experiment or experience. What they do not realize is that they are committing the double sin, which is unforgiveable. First they assume the Universe cannot explain itself, but they can explain the Universe. And second, in so doing, they eliminate from their own potential the possibility of communicating directly with the Universe. What I hope to introduce you to is the benefit of communicating directly with the Universe; no intermediaries required.

The source of the energy that moves particles through space is space itself. Let's get a little more comfortable with this idea. Let's use our imagination. We're going to take an idea . . . space . . . #1 . . . and give it form; visualize it . . . draw the picture . . . #2 . . . in the invisible Mind . . . #1 . . . let it become a more reasonable cosmology . . . #3 . . . so that we can manipulate the energy of space. And where does all this energy come from? Not from a Big Bang. That's too complicated. But quite simply, it comes from the other side of space. We've already mentioned the fabric of space. What is the fabric? It is like a garment. On one side of the garment, on the inside, there is all light. On this side, the outside, where the physical Universe is; where effect is; there is all form.

The light comes to us through the seam in space. Imagine the seam is like a zipper. When the zipper of space is opened, light comes through to this side where it becomes all the material form of the Universe. And what is this seam? Is it a zipper? No, it is much simpler than that. It is the wave. The wave that is associated with every particle (and every particle has its own wave) is the seam in the fabric of space that creates the particle and provides the power which moves it through space. Couple this one source of all light—namely "the other side of space"—with our own intelligence, and you can imagine that this Universe is relational, which means we get to create our own destinies in this one vast space.

I believe we have the potential to be the pride of the Universe. Maybe if we all had to pull together to save this planet—our home—from some sort of catastrophe; or maybe if we could control the global weather by tuning our collective consciousness into the higher consciousness of our Mother Earth; if we could find some way to unite all the people, we could control the energy of the planet, then the energy of our solar system, and one day, the energy of our galaxy. To do this, we are taking the steps to learn what space is, because all these activities are happening in the one space that fills this entire Universe.

There is only one space. It is the source of the energy that moves particles through space. Everything in the Universe consists of particles. Therefore space has created everything in the Universe. "Everything" includes intelligence. That means space is the source of all intelligence and it is intelligent. You know from your own experience that intelligence responds to intelligence. This fact is what gives me the courage to help others heal from life-threatening illnesses; because I know it is not just me alone who is trying to help. I do not rely upon my own powers. When I deal with an illness, whether my own or someone else's, I cohere the random background fluctuation of the energy of the space that contains the person who needs to heal. Then the same energy that builds the galaxies is right there, in our immediate space, rebuilding the cells in the human body because the same space that contains us is contained within us. And we have the authority to cohere the background fluctuation of the space that contains us because there is only one space and it is us.

Third Chapter

DEGENERATIVE ILLNESS

A degenerative illness is one that continues to get worse over time. There are several types to choose from as we investigate the cause of all degenerative illnesses. Let us begin by assuming the energy matrix of the illness is smaller than the power source of the space that contains it. By this I mean, the individual who is suffering from the illness occupies a very special portion of space which we identify as that person's life. There is one energy source that acts as a backdrop for all the events of that person's life, and it is the source of all the energy in that life. And it is intelligent. Much more intelligent than people realize, actually. The background energy of space—the Universe—is smart enough to give to each and every one of us exactly what we expect of it. If we act like it is dumb space because there is nothing there, then for us, it is dumb space. Things just seem to happen to us "by accident;" under the umbrella of the self-wish described by "you just never know." But, once we know the truth about the energy that contains us; that it is intelligent and can respond to our intelligence, then the forces of the Universe act in our behalf. The fact is, this relational Universe gives back to us exactly as we expect it to. We can think of this as the energy version of Newton's third law, which states that for every action there is an equal and opposite reaction. This law works very precisely for matter and for energy.

Let us assume the physical world is the world of effect, as we have shown earlier. Then it would follow that since Newton's laws work on the physical plane, it must be that they also work on the causal level which precedes it. Remember, for every effect there is a prior cause. For Newton's Laws to work so precisely on the level of effect, it must be that they work just as precisely on the level of cause. And so it is that for our bodies to function on the perfect and flawless level for which they were designed, our minds and our thinking capacities must occupy some level of perfection which is mirrored in the physical. And this brings us back to the belief we all have that matter is solid and space is empty. If this is not a true statement; if it is just a couple of false assumptions shared by everybody; then it must be that we are all missing something upstairs. One of the effects of this collective flaw in thinking is that it allows for the random occurrence of degenerative illness. We know this to be true for cancer, for instance, whose attack mode is often described as random. This happens because our collective thinking is flawed. But we can overcome that by understanding how to cohere the random background fluctuation of the energy of the space that contains our bodies; putting them in tune with the perfect energy flow.

There is an innate mechanism in our thinking which is carried through from generation to generation by the subliminal consciousness of the species, otherwise known as the race mind. It is linked to survival, therefore its harmonic subsumes all others. By that I mean, all the other functions of life follow the survival instinct in terms of importance. That is why the accuracy of our perception is so important. In the finite world of three-dimensionality, the human animal is like all other animals. If it sees a threat, it runs away or it vanquishes the foe. It has to see cause and effect within the observable action so it can make survival decisions. That is the sort of Neanderthal logic that is buried deep within our human universal mind. In order to sustain that logic, we do whatever it takes to avoid the simple truth about our own existence; which is that there is one life force that is common to us all. We are all variations of the One but we are different and separate on the level of effect. We are united on the causal level because there is only one space in the Universe and we are all contained within this one space. Such a non three-dimensional way of thinking cannot be allowed into the three-dimensional octagon of battle because it doesn't stand a chance. It's not physical. It has to lose in a fight. Therefore it must be forbidden. But that is the old way of thinking.

We have established that there is intelligence in the Universe and from the Laws of Space let's consider the Third Law of Supradimensionality, which is the Law of Infinite Exchange. It is the energy counterpart to Newton's Third Law and it establishes the framework of this relational Universe, which is constantly giving back to us what we expect of it. Because we think it is finite, it acts like it is finite. "But wait a minute," you're thinking. "If it is a Universe of principles, it cannot be capricious. And that giving back in accordance with what we expect sounds whimsical." But I must tell you, it is not. This is a lawful Universe. But it turns out that ignorance of the law does not leave us guiltless when we break the law. That would be capricious. Instead, the balance of the Universe is always accurate. What do we get in return for thinking the Universe is finite? You could say we get nothing for that false belief. But what do we get for believing in the infinite Universe?

We get unlimited access to all the intelligence in an infinite space when we assume it is intelligent. We get what we pay for with our spirit of belief. If we think it's zero . . . we get zero. If we think it's everything . . . we get everything. Whatever we believe our relationship with the surrounding space to be, that is what our relationship with the surrounding space is.

These ideas bring us to a reasonable consideration of how degenerative illness works. There is a constant exchange of information between us as individuals and the Universe in which we live; more specifically, between the individual and the space that contains that person. When our belief is immature or incorrect, we restrict the flow of information. Life is like any course you take in college. If "Life" were your major, there would be introductory classes on manners and social etiquette; followed by 200 level classes in dialogue styles or dress codes; perhaps 300 level courses on how to be rich and famous. What you find as you go through the lessons of increasing scope and complexity and, usually, increasing difficulty as well, is that as your understanding of the topic broadens, the conduit of information flow between you and the universal source also widens, which allows more information to flow. On the other hand, as any portion of information flow is diverted, usually because of ignorance, there is a precisely concomitant darkening of the light that is in transit. This is another way of describing the Law of Infinite Exchange.

I keep harping on these Laws of Supradimensionality because we begin the process of acceptance of principles by voicing them and giving form to what is otherwise formless. What we think about we bring about. That is to say, thoughts do become things. We think so fast and so many pictures come across the video screen of our own personal awareness of life, that it would be cumbersome and unmanageable if everything we ever thought about became our own personal reality. So the process is narrowed by the contingencies of life. We may want the $100,000 sports car, but we have to settle for the sedan we bought for one tenth of that amount. However our own self-estimation is derived, it is invariably what we expect, that manifests for us. The single most important question then becomes . . . Do we really know what we expect of ourselves, and therefore what we are constantly creating?

The answer to that question is, no. We do not know what we are really creating. "But how can that be?" you ask. "What do you mean when you say, we do not really know what we are creating? Why, this is 21st century America. We send men to the moon and robots to mars. Surely we know what we are thinking!" And you continue. "And we are smart enough to partially control the forces that manifest in our own individual lifestyles and our nation's dominant influence on the affairs of the globe." But you see, that's all the big stuff; the stuff that's got everybody's attention. There are also some elements in the foundations of our individual and collective mentalities that are ignored because we are unaware of them. That lack of awareness is the darkness I referred to above, which lessens the flow of information. ALL DEGENERATIVE ILLNESS IS CAUSED BY ENERGY DEFFICIENCY. The effect is similar to having something growing in the gas tank of your car. Every time you fill up, the tank holds less and less fuel, until one day, it holds so little the car won't start.

What is increasing is entropy; wasted energy. On the fundamental level, it is not so much a matter of what is growing inside the body of the person suffering from a degenerative disease (though in the case of the Big "C", it is a perfect metaphor); it is more a matter of what portion of energy is not able to keep pace with the demands of the system. All of our life we are growing. It is not like the growth of some thing which interferes with the flow because the growth would be the effect; the result of the diminished flow. The cause is in the unseen world of energy. And in that world all energy is intelligent. Its first act is to create. That which it creates evolves. All creation evolves. It is never static. It evolves through the mechanism of infinite exchange. And those are the three Laws of Space:

Creation—Evolution—Infinite Exchange.

If a person's thinking is locked into the popular belief of the Big Bang Theory as the defining statement about our Universe, then that person's energy flow is incomplete. A portion of the energy that is available for our body's successful health and well being is not being used because we have to think our way out of the belief that we can observe cause on a three-dimensional level. That is how degenerative illness is allowed to progress: by not being energized out of our system because a certain lack of awareness on our part prevents all the informational energy from flowing through to us. But what would happen if the floodgates could be opened so the energy could flow?

When more energy is added to a system, particles are accelerated. They move faster. They can move smarter, too, if we connect them with the source. That is to say, recognize intelligence and it responds to our intelligence. I am talking about the intelligent energy of space, which contains us and is contained within us. When we start to rev up from the truth of being, we are tuning into the same force that builds the galaxies. That almost sounds too big to handle, but the truth is, we are all cosmic beings, yearning to seek out the truth from the elegant space which is the one source of all power in the Universe. We are not trained to think this way, but if we don't learn to think this way, we get sick. Open the floodgate just a crack. Intelligent energy—information—begins to flow and the illness begins to heal.

What we are going to do is replace the degenerative illness with regenerative healing. We'll use a technique called sympathetic vibration.

First, recognize that the energy sources of the Universe are vast; but as big as they are, they are available for you and me to use in the powering of every day. Did you ever walk through a forest? Every tree you walk past is taking in your CO^2 exhale, processing it through millions of cells of tree trunk and limbs and leaves and giving you back oxygen for your next breath. The winds that blow through the trees and across the planes are headed right toward your lungs, to supply you with the fuel for life itself. Our link to life is immediate and intense. Everything is energy and energy moves through waves. When you and I link our energies with that of the Universe, through the power and intelligence of the space that contains us, we synchronize. Sympathetic vibration is the vibration in one object that is caused by a similar vibration in another object. We think of ourselves as living objects. When we are close in thought, the healing vibration is absorbed by you and the energy of rejuvenation lifts up those deficiencies. Begin the process of returning from deficiency to the efficiency and perfection you were born with.

Sometimes it seems like another mind is thinking for us. Or, in the case of degenerative illness, that other mind is thinking against us, as it were. Rampant growth, for instance, in the case of the Big "C", is a function of some variable that is beyond our control. It is cellular growth, so it is under some control; but it is not under our own self-control. At least, we currently don't think it is. But then, we don't currently think our collective belief in the emptiness of space is of any consequence, either. But it is precisely that perception-based assumption that robs us of our link with the infinite. The mind of the Universe is much bigger than our own mind, though we are part of that larger consciousness, and receive from it in exact accordance with how we believe into it. Assuming it doesn't exist does not harm it at all. But that human assumption allows the universal fields to function through us at random. To heal ourselves of rampant growth we take control of the energy behind that growth. We cohere the random background fluctuation of the energy of the space that contains that tumor. The body heals itself.

Strange as it may sound, especially to the one who is suffering, there is a hidden blessing in degenerative illness. Since all deficiencies of energy can be remedied by simply adding more energy, we learn how to heal ourselves by becoming more tuned in to the intelligent energy of space of which we are composed. We learn to synchronize our energy with the energy of the Universe. Since great minds think alike, we link our mind to the Mind that built the stars. There is such a Mind and we are always given the opportunity to unite with it through our own volition. We always have the choice of thinking like the skeptics and tuning out the vast energy resources of the Universe, or taking a leap of faith just because part of everything is a lot more than all of nothing. As soon as you say, with understanding and conviction, "The Universe is infinite and I want to know what it knows," you deny the Big Bang Theory.

In so doing, you stop it from closing off the flow of healing energy through you. That double negative is a positive. Follow that by affirming the Universe has always existed, which opens the flow of healing energy and, sure enough, you begin to heal. In time you will completely defeat the degenerative illness and once you have done that, you can help others to heal. Go out of your way to protect your family and your friends. You have the power to do so through the One Power which permeates all space and lives inside of you. All of these acts of love strengthen our relationship with the Force of the Universe. Remember our goal, which is to control the energy of our planet. We are one step closer when we triumph over illness.

Fourth Chapter

BELIEF

I believe we can stop the suffering by teaching the truth. Remember, words of truth are the violin strings of space which vibrate the creative power of the Universe. Which suffering can we stop with the words of truth? To give you an idea of the scope of what we're talking about, let me tell you that the idea came to me right after I heard the news about the beauty parlor killings in California. That wasn't too long after the massacre in Norway. What causes random acts of violence? How do these relate to energy deficiency? That is actually the single most important question in the world right now. If we have to rely upon our own limited, finite resources to solve all these energy deficiency issues, where is all the power going to come from? We're already having brown-outs. That's like imagining the link with the universal supply can be bought with money. If it could, where would we get all the funding from? We already have cities going bankrupt. Actually, there's not enough cash in the world to purchase the key to the cosmos—to the universal intelligence that pervades all of space. It can only be bought with faith. That is the key used to open the valve to unlimited energy resources; sources big enough to build galaxies. Faith is knowing, from your own observation and experience, that space is everywhere present. What could be simpler than that? Just look around. Then add to this observation the understanding that space is the way we humans perceive energy. The missing wavelength discovered by Qc^3 is the link between California, Norway and energy deficiencies in general. It is the source of all intelligence. As soon as we begin to consider the unifying factor of the energy and intelligence of space, which is infinite, we begin to heal the mentality of the planet from its propensity toward random behavior.

We all believe in science today because the practitioners of the scientific discipline are able to make things happen. So we will continue to identify life-changing principles in the cause and effect vocabulary, remembering the importance of thinking in three's. With cause as one and effect as three, that leaves number two—the action—to be discussed. That action is belief. It is where the rubber meets the road. The idea acquires the substance of form through the action of belief. This is taken down to the atomic level where the particles ride on their appropriate waves of energy as those waves cut the seam in the fabric of the one space to create form. There is design and direction behind the motion of every particle that moves through space because there is only one space and on the other side of the fabric that we perceive as empty, there is all light. So that one source of light is the blip, if you will, in the fabric of space that becomes the photon and every other particle contained in space. That means there is one source of all light that is manufacturing those particles and sending them on a destiny which we participate in; either because we have summoned them forth through belief or someone else has. The second law is Infinite Exchange and it is transacted through belief.

Our capacity to believe is worth considering because, as you can see, its position is fundamental in the creative process. Everything that we see has been created by that which we do not see. And the action that brought that demonstration about is belief; or no belief at all, in which case the creation was the result of random fluctuation. We have to always include that as a sort of creative disclaimer. In space, matter is always being created. If the random fluctuation is cohered by our belief it creates something. If it is not cohered by our belief it creates at random. It is said that what mind can conceive, man can achieve. A more thorough version of that dictum states, what mind can conceive and believe, man can achieve. We cannot leave the second step out of this process or we will get nothing but random fluctuation. So what is belief? It's not a thing, though belief is what creates the thing. It is a concept, which means it exists in the unseen world of thought or energy, where all cause exists. But it is the process that moves the particles into the form they end up occupying. Therefore it must be a power because it is actually doing work.

It is the power of thought.

But how do we measure it? You know that once we figure out how to measure it we will be able to exploit it as a source of power. That's what we've done with every other source of power we have discovered; i.e., figure out how to make it useful, package it and sell it.

So that is just what we are going to do with the power of thought or the power of space; i.e., figure out how to make it useful, package it and sell it. And what could be so useful that it would be of value to society? How about this? Since we don't teach the truth about space now and we live in a world where random acts of violence make the headlines every day, let us suppose that by teaching the truth about space, we can change human behavior. In fact, we can be totally affirmative about the result because we know, with complete certainty, that when we change anything on the causal level, the effect has to change, too. When we change the way we think about space, something on the effect level changes as a consequence. If you will flow with me now, Fellow Traveler, in your imagination, to the other side, we can ease the suffering. First to the other side of space which unites all minds; then to the other side of the event to heal it before it happens. There is energy available to us that we are not using right now. We need to rev it up. We need to activate it.

Sounds like a pretty big job; maybe so big we shouldn't even attempt it. How in the world are we going to change the attitude of an entire planet? The answer is, one believer at a time. The beauty of this one-at-a-time approach is this: when we're using space as the backdrop for our dialogue, it's like baking a loaf of bread. Space is the flour. Belief is the yeast. When the yeast is activated, it affects the flour in the entire loaf. And once it's been affected, it cannot be un-affected. The portion of the flour that you represent, once raised, becomes a different product; different from what it was before it was affected by the activated yeast and certainly different from the same amount of flour to which no yeast was added. But who adds the yeast?

Here's where the Hundredth Monkey comes in. Again, from Wikipedia: "the hundredth monkey effect is a supposed phenomenon in which a learned behavior spreads rapidly from one group of monkeys to all related monkeys once a critical number of initiates is reached. By generalization it means the instantaneous spreading of an idea or ability to the remainder of a population once a certain portion of that population has heard of the idea or learned the new ability by some unknown process currently beyond the scope of science."

My Dear Fellow Traveler, that unknown process is now known with this writing; and while it is beyond the scope of the science that currently runs this planet, it is well within the scope of the New Science of Space, which easily explains it this way. There is one space which contains all the monkeys so the minds of all the monkeys are connected, because the space that contains them is identical with the space that is contained within them and it is intelligent. It is interesting to note that the phenomenon occurs within a population. In the hundredth monkey, as it was first observed in 1952, these were macaques; monkeys that inhabited the Japanese Island of Koshima. In our case it is the human population, which currently occupies the planet Earth. There is some critical number of initiates that must be reached before the trait is passed from the group to the whole population. That critical number is 5% of the population and it begins with you and me.

It is uniquely challenging for us to note also how Wikipedia refers to the hundredth monkey effect as a "supposed phenomenon" and goes on to explain how the concept was accepted and put forth by some, but discredited and therefore disproved by others; namely, the skeptics. But that is precisely how the Universe works when it comes to belief. To those who believe in the power of humanity to survive and advance as a species on this beloved planet, there will be survival and we will advance. To those who claim they cannot believe in the existence of what they cannot see, in keeping with what they know to be true as animals; they will just be coming along for the ride, as it were. (Check it out: there are about two billion cars on the planet and everybody wants to drive one, but there are small groups of people out there, even in the most advanced societies, who still rely on a horse and buggy to get around. And there's a lot of folks in the more advanced groups who are jealous of the simple lives those horse-and-buggy people enjoy and wish they could be like them.) And so it is with belief. For some, the tree in the forest has fallen; for others, it still stands.

Fifth Chapter

COLLECTIVE CONSCIOUSNESS

What we are going to focus on in this chapter is what I hope you will come to know as your responsibility to the Universe. A lot of people like to think like the guy who said, "I didn't ask to be born into this life, so . . . . whatever I do is whatever I do!" (plus whatever else he said after that). The point is, because we forget about the origin of consciousness, we forget to take responsibility for our own lives and our own behavior. It should come as no surprise that this behavior, this lack of accountability, is actually supported by the philosophy of three-dimensionality; which is the system of thinking we are all born into and therefore accept without question for the rest of our lives. But this is unlawful, because it implies that ignorance of the law is an excuse for breaking it, which it is not. It is criminal to promote a criminal act. And not being responsible for your own behavior is against the cosmic law. In this relational Universe, it is your duty to know who you are and at least what minimal level of accepted behavior is expected of you. Expected by whom? How about this: expected by the planet Earth upon which you dwell.

Expected by whom? How about this: expected by your fellow inhabitants of the planet. After giving it some consideration, we realize that every society organizes itself by accepting codes of behavior and living up to them. Some societies seem to be more successful than others, but that success must be measured against the backdrop of the planet's well being. It is said that if we wish to mold the behavior of society, we first have to shape our own behavior to fit that mold to set the example. Similarly, if we want to control the energy of our planet, first we have to control the energy of our society and to do that, we must begin right in our own home to modify the energy of our selves and the energy exchange we have with the ones we love. We are all part of the collective intelligence that encompasses this globe. Before we were societies we were nomads and cave dwellers. We discovered that by joining forces with others we could live better, safer lives. And we evolved. We built nations and churches because we found ways to live better, safer lives. Soon we will all realize that it is time for all societies to modify their behavior to handle the changes happening right now to our planet's atmosphere. At last! We are given the one great cause that unites us all for its potential achievement, and the means of communication by which to do it.

Life imitates art. This is a very nice philosophical description of the creative process that is behind all manifestation in this so-called "physical" Universe. We must assume that by "life" we mean this human existence on the planet Earth. By "art" we mean the creative process. All art, whether it be music or poetry; comic strips or Gauguin; it is a creative act. We have seen that all matter and hence, all material creations, owe their existence to the energies that move the particles that form the matter. So, energy precedes form. Intelligent energy is the motive force behind the creation we live with here in the material world; i.e., the world of demonstration. My favorite example of this process is the work of Jules Verne. Eight predictions by that author became reality, but the one I like the most is his 1865 publication of "Voyage To The Moon." The art form describes a capsule housing men being blasted off to the moon. A hundred years later we imitated that art by sending the Apollo 11 capsule to the moon. Thank You, Neil Armstrong, for "One small step for a man, one giant leap for humanity."

Life imitates art. For each and every one of us, life is the representation in physical form of the art work we wake up and perform every day and dream about at night when we go to sleep. So what are we actually creating? The answer to this question is readily available in the history of humanity, which keeps track of what we have done here on the planet Earth; and in an unbiased look at current events, to see what we are creating in the present moment. It looks like we have made some progress in some things, like medicine and architecture; surely in farming and pharmacy. But in some ways we haven't progressed at all. War, for instance, is still an acceptable mode of behavior among societies. Starvation and hunger are pitied, but never remedied. Everyone on the planet wakes up each day expecting life to be just the way it appears to be, with solid objects moving around in empty space. This fraudulent interpretation of our perception is verified, supposedly, by everything we do and that misinterpretation actually invented the finite Universe and uses it to substantiate its own doctrine. So we all believe in a lie about the Universe and remember: thoughts are things. Is this what is holding us back as a species?

What are we thinking? Have you seen the law of balance at work in the Universe? We call it karma. In this relational Universe we are all participating in the flow of life. That life force is an energy of give and take. We have all seen how when we give, others give back to us. And we have seen that when we take something which we have no right to possess, it isn't long before something of equal or greater value is taken from us. This would be the moral rendition of Newton's Third Law: for every action there is an equal and opposite reaction. Be careful with this law because it sounds like it is the action which creates the response. But there is also the act of no action. Not acting when we are supposed to act is just as bad as acting wrongly. Now we need to back up this entire dialogue one step to the level of thought, which precedes all form. If you do not choose to take the time and trouble to figure out precisely what your relationship with the Universe is, then you are at fault. Don't you want to protect yourself and your family? From degenerative illness? Then . . . know thyself!

The good news is, you don't have to know everything. The tendency is to think that because we are individuals; because our behavior is what counts in our own lives; we have to "tow the mark," as it were, if we expect to be successful and well rewarded. But this conviction of sole responsibility is misplaced in only one direction; namely, don't leave the intelligent energy of space out of your own personal equation for life. Remember that the space that contains you is the same as the space that is contained within you. Please recall that there is only one space in the Universe. If you can accept the simple truth that the source of the energy that moves particles through space is space itself, then you realize that there is only one source of all energy and all intelligence and it is the space that contains everybody. If that is the truth, and you don't believe it or respect it, then you close the door to that energy source because it is intelligent, just like you are, and it always responds to us according to our belief. It always substantiates our convictions. If you think it is nothing, then for you it is nothing.

And that's too bad. Too bad for you and too bad for those you love. Think about the possibility that the destiny of the planet is to become what some of the science fiction writers have predicted. I'm not talking about the planet ravaged by war and destroyed by human negligence. I'm thinking of the planet that has learned to control its own energy so it can travel through the solar system, developing planets and moons so our burgeoning population has places to go and work to do. I'm thinking of the planet that feeds all of its inhabitants well and controls its climate forces so we can all live richly rewarding lives. And this is all achievable if we join our thoughts together for the good of the globe, with the knowledge that our minds are connected to the human mental grid we call collective consciousness. There is a unifying principle which the various religions teach and it is good to know for sure, thanks to the proof of science, that there is oneness behind all variety. That way we can also imagine a space that actually cares about who we are and how we are doing. We are all connected in space, by space. It is helping us to achieve perfection.

The QcCubed information-based therapy promotes active unification between you and the space that contains you. Imagine the space that surrounds you and penetrates through you, in the atomic structure that makes up every cell of your body. Imagine that you use the directive force of your own individuality; the **I AM** that describes you and who you are; to cohere the random background fluctuation of the energy of the space that contains you. **But the space that contains you is intelligent.** The intelligent energy of space created everything in the Universe. Imagine it is like a universal library we all have access to. Realize that it helps us if we ask it to. I prefer to think of it as the Mind of the Universe. Our own brains are like the individual PC's that are all connected to that big "server in the sky" that is like the World Wide Web of consciousness. With eight billion brains on the globe, that's a lot of storage capacity.

To change our behavior we have to change the way we think. That is not an easy task. Once we develop a functioning lifestyle, we are not quick to alter our behavior. I'm hoping the idea of changing our collective consciousness becomes a fad; something everybody wants to do once they see how satisfying it can be to help one another and how important it is for the destiny of our planet. To change the way we think it is important to know how we think. Behind the behavior we demonstrate there are two formative thought process groups. Each group represents a behavioral paradigm that is antecedent to what we actually do. One we shall link to subliminal behavior, the other to cognized behavior. The unity between them, like the pool of profoundly deep clear-blue water upon which they float, we shall come to know as our Universal Identity. We are the pride and joy of the Universe but we are not taught that. Now that we know what space is, we are ready to learn the truth, the whole truth, and nothing but the truth.

Sixth Chapter

PERCEPTION

This Universe was created for us. Let me ask you a trick question. Maybe you've heard it before. If a tree falls in the forest and there's no-one there to see it fall, did it really fall? The answer is: that's up to you. If you, in your mind, want to see it still standing, then it did not fall. I can tell you it fell for me, because I am the center of my own Universe and I see it on the ground. You are the center of your own Universe. If it is still standing in your world, then it did not fall. It fell in my Universe, but not in yours. How can that be? We all live in the same Universe. How can there be two different answers? How can it be, and not be? The answer is: our participation in the Universe is what makes it real for us. This entire world was created for us. All of space is a sea of intelligent energy that has existed forever. It has no beginning and it has no end. We, in our desire to make the events of life intelligible and manageable, have transferred the limits of our perception to the philosophy of life we use to transact our daily existence.

When I looked up "space" many years ago in a standard reference encyclopedia, it was defined as "the unlimited three-dimensional expanse" that filled the Universe and contained all the actions performed therein. But I thought, how can we use three dimensions (length, width and height or length, width and depth) to define the objects contained in space and use the same three dimensions to define the space which contains the objects? After further consideration, it became apparent that there were certain assumptions we took for granted, but which were not necessarily supported by the facts.

Think about our situation. We are all born into a system of thinking that is at work when we arrive here on the planet and we learn that system. But what is it? How come nobody ever talked about it? What was the system we learned? Was it even a system? After giving the matter thirty years of deliberation, it became apparent to me that there is indeed a certain philosophy of life that is everywhere functioning when we arrive here and it is so universally accepted and so common that it never even occurs to us to question it or think about it. I could not find reference to its use anywhere, so I decided to call it "Three-Dimensionality, the Philosophy of the Obvious". Its most fundamental premise is the observation—no, more: the self-fulfilling prophesy—that **life is the way it appears to be.** Meaning, among other things, that objects are three-dimensional and so is space. And the first corollary of the learned system must be the observation that matter is solid and space is empty. All the subsequent laws and principles that we assemble to define our existence will be based upon these facts which are proven by our observation of them. You already know the problem here. Even though these facts are used by science they are not substantiated by scientific understanding. Nonetheless, we go on to use the solidity of matter and the emptiness of space to define our Universe, ignoring the consequence of believing so wholeheartedly in a way of thinking which cannot be true. So what should we do? There is only one thing we can do. That is, think deeper.

To begin with, don't necessarily believe what you see, learn to see what you believe. This material world is the world of effect. We are told in elementary science class that we see objects because light bouncing off them or light being given off by them travels to our eyes carrying the information about those objects to our sensory mechanism. Once we are aware of an object we begin immediately to share information with that other accumulation of matter through what can be envisioned as the process of infinite exchange. The infinite part of that exchange is not necessarily that we will be involved with that other accumulation of matter forever; but rather, the infinite dimension of that otherwise linearly measured event is the nature of the energy exchange itself. It is not measured by matter because it creates matter and it is intelligent, so it can be influenced. However, if it is to be influenced directly that cannot be accomplished by manipulating it on the level of effect. If we change the material part of the process, that new demonstration influences the next causal action, but we changed the effect by first deciding on the causal level to do so. Sometimes it looks like the effect changed the cause, but it didn't because there is always some prior action on the causal level. That's just the way things work around here.

We have been taught to believe what we see, but the fact is, we are really always seeing what we believe. When the light "bounces" off the object and travels to our eyes, what it has really done is better described as an absorption/ emission process. Light doesn't just "bounce" off of anything. Bouncing would be like billiard balls colliding with each other or with the cushions of the pool table, where angle of incidence equals angle of reflection. However, on the particle level, it is not reasonable to assume that one particle, like a photon, actually collides with another particle, like they do in the Large Hadron Collider(LHC) in Switzerland. Particles don't collide in everyday life.

When a photon "strikes" an object, like a photovoltaic cell or a mirror, it seems to be absorbed by the PV cell and reflected, or "bounced off," by the mirror. In fact, it is always absorbed by whatever it collides with, and given off again. In the case of the PV cell, it is absorbed by the atomic structure it encounters. More energy in the atom causes electrons to jump to other orbits and in so doing, photons are given off which travel back to, in this case, our eyes; because we are involved in this particular observation process. So the light that struck the PV cell got absorbed and given off again. Some of it is given off as heat, causing the device to warm up. Some of it is given off as an increase in the flow of electrons, which creates the electricity we use or store in batteries. And some of it is given off as the information that travels back to our eyes and tells us what it looks like, how big it is, what color it is, etc. In other words, part of the information that is given off is the object's identity.

The light that "bounced" off the mirror did almost the same thing. Because so much of the light was reflected, the mirror absorbed less heat. It did not increase the flow of electrons in a suitable current flow mechanism, so no electricity was generated. A portion of the reflected energy brought the identity of the mirror back to us: how it looked; how big it was; its shape, etc. But at no point did particles actually collide with other particles. If you want to do that you have to build a multi-billion dollar particle accelerator and aim particles at each other using powerful electromagnetic force fields.

The point is this: we don't see an object because light "bounced" off that object and traveled to our eyes. That bouncing effect is probably still being taught in elementary science but it is entirely false. We see an object because light has been absorbed by the surface molecules of that object and this added energy accelerates other particles so light can be given off by the object. It is this light, which carries information about the object's identity, which travels to our eyes. We don't just see a reflection of the object, which is what "bouncing" implies. We see the thing itself; not photons "bounced" off the object but photons from the object itself. Remember: in a reflection, everything is reversed. If you write your name on a piece of paper and hold it up to the mirror, you see your name is backwards; the letters are backwards. But when you and I look at each other, we are not seeing the reverse of each other because it is not reflected light, it is emitted light. And where does that light come from? It comes from where all light comes from, and all light is on the other side of the fabric of space. And it is intelligent. So that is where we influence what is being created. Where all light is, there is also all intelligence. The more we understand that what we believe is what we see, the more we influence the energy that creates the effect on the material plane.

This process is going on whether we realize it or not. It invites our participation but it does not need our participation. It has it anyway! We are alive. The Universe created us and it is living through us, as us. If we never figure this out, we go through our lives doing the best we can to survive, taking the shots from a life of random occurrence and giving life back a few shots of our own. But it is not until we know how things really work, that we can make a real difference, not just in or own lives, but also, and more importantly, in the lives of those we love.

What is perception? Before it is anything else, it is a choice. Please allow me to explain. For each and every one of us, in a typical day there is quite a wide variety of incoming information. We typically don't try to absorb everything because we don't need to and we really don't want to. Too much information makes us uncomfortable, so we tune it out. What we process is what we focus our attention on. Once we have decided to allow stimuli to enter our system, the information is then "translated" by as mechanism called selective perception. After the information has been translated into our own personal code, it gets stored. The question that begs to be asked is this. If perception is a choice, who is doing the choosing in instances where we are not consciously aware of the incoming stimuli? Imagine, for instance, that we are in the mall and Christmas is just a few weeks away. It's packed with shoppers. A child starts crying. Some people hear and do nothing. Some hear and respond. Some who are standing right next to those who hear the crying, don't hear anything. They are pre—occupied. They just saw what they have been looking for. They don't hear anything because all of their attention is focused elsewhere.

The child's crying means nothing to them so it gets no attention at all. But the same sound is traveling with the same intensity to the person standing right next to them. One hears it, one does not. Neither planned on hearing a child cry. One heard, one did not. Same stimulus. Both have functioning auditory mechanisms. On a subliminal level, one chose to hear the child drying, one chose to not hear the child crying. So we see that there is the conscious level of choice and the subconscious level of choice.

What does all this have to do with healing? Well, let me ask you a related question. Who drives a car better, the person who knows how everything works or the person who doesn't know how anything works? You know the answer to that question from your own experience. When we first got our license, we were clumsy behind the wheel compared to how we could maneuver after a couple of months of practice. If we drive professionally, we become very good at driving. Like they say, practice makes perfect. Experience is what teaches better than anything else. And knowing how any system or device works makes using it that much more elegant and effective.

The same is true with healing. Knowing how the healing process works will make the teaching of healing that much more effective. Once a person learns how to heal, either their own sickness or that of another, they can teach others. And that is the goal of the QcCubed Information-Based Therapy. I have healed myself and others and they will go on now to teach others how to heal. We are focusing our attention on degenerative illness because it relates to a flaw in our thinking which is holding our species back from continuing to evolve. It is one thing to read about how to heal and another thing to actually heal, especially from your degenerative illness. A lot of folks are given no hope at all of ever being cured of their disease, so they have to settle for "quality of life" issues and just do the best they can to get by. But this information-based therapy works! As you shall see.

Let me tell you now about my interpretation of the Huna philosophy of life because it introduces us to the three personalities we each have. Not just one personality, but three! We consist of a lower self, a middle self, and an upper self. Notice again that we are thinking in three's, so we are off to a good start. By opening our minds up to the possibility that we may each have three selves to deal with, we open up potential lines of communication with those selves. Keeping it as simple as possible, let us just say that we have the subconscious mind ruling the lower self, the conscious mind leading the middle self, and the Super consciousness, the Mind of the Universe, working with us and for us through our higher self.

With three levels of consciousness there come three levels of awareness. Each is qualified to choose, in our behalf, what information to allow into the system and what level of response to authorize. If you want to heal from your degenerative illness the first thing you have to know is that others have healed from the same disease and if they have not, you will be the first to accomplish this miracle. Because the energy that will overpower your illness is intelligent, you don't have to know how it will heal you, all you have to know is how to choose to have it work in your behalf. With your conscious mind you recognize the three selves and authorize the Mind of the Universe to work through all levels of your personality; to just let go of all resistance to truth you may have learned in your youth or built up over the years, so the healing energy of space can cohere the random fluctuation of the energy of the space that contains you and your body. Then thank the Universe for healing you. Listen, then, over the weeks and months and years that follow, to the voice within your own mind for advice and direction. Your Higher Self will honor the request of the Middle Self and the vibration of the Universe will tune into the Lower Self and synchronize it. And you will be healed.

What is perception? It is interaction. You perceive on three levels. You see, hear and feel things you are not consciously aware of. The voice of the Lower Self wants us to live up to the limitation it learned as a child. The voice of the Upper Self wants us to become Masters of the Universe. The Middle Self has the voice of reason with which to speak words of truth into the demonstration of healing from all forms of degenerative illness. Now that you know how much energy is available to you, please speak the healing word for yourself with confidence. And you will be healed. Then you will go on to heal a loved one. And the chain reaction has begun.

Seventh Chapter

PRINCIPLE

The Universe works in 3's. If you don't think that matters to you, then remain non-committal and uninvolved for a few more years while the events of the planet upon which you dwell become a little more severe and life threatening. At some point, somebody who cares about how the Universe works and who cares about you, will engage you in the healing dialogue and make you realize that the first principle of the Universe is that we are responsible for its existence, as it relates to us. The rulers of the Universe are those who truly understand how it works. And it works in 3's.

In the philosophical vocabulary we identify mind as the source, thought as the action, thing as that which is created. In the scientific vocabulary we recognize the cause, action and effect. In the supradimensional vocabulary, from which issues forth the science of space, we imagine the energy of space being cohered, through belief, into the physical reality we live in. Simultaneously, we live in the climate of our beloved Mother Earth.

The last ten years have been the warmest decade on this planet since we started keeping records. We are involved in climate change. Some people think it is our fault that the planet is getting warmer and some people think it is not our fault because the planet goes through cyclical changes anyway. Then there are some who realize that it doesn't matter whose fault it is. What matters is that we do something about it. But what can we do? It's already too late according to some. The environmental damage has already been done. If we stop all carbon dioxide emissions on the planet tomorrow, we're still going to keep warming up for decades just based upon the CO_2 we've already belched into our atmosphere. And believe me, we're not going to stop pumping carbon emissions into the air we breathe for a long time. The "high" of becoming successful and being able to enjoy the finer things in life, like cars and consumer goods and yachts and fine cuisine; this "high" is tough to resist when it comes your way. The emerging middle class in the developing nations is pressuring manufacturers to produce more, not less. So we can all realistically expect our environment to continue to absorb not just more CO_2, but a lot more CO_2!

It looks like the problem is that we can't stop poisoning the atmosphere of our planet because we can't stop wanting to live better lives. Because we want a house and a car and some nice furniture and some nice clothes, we have to manufacture more goods and that's causing more carbon emissions. So, one might suggest that we figure out how to satisfy ourselves with less. But that's not going to happen. If we can afford to live better, we have a right to live better, more productive lives. And we are right to want better, more productive lives for our loved ones, too. Wanting less, then, is not the solution to the problem. So what is the problem?

The answer must begin with getting the right idea about the space that contains us and our planet. Do we all currently realize that it is also the same space that is contained within each and every one of us? Of course not. Do we know what the energy of our planet is? Do we even know what energy is? We don't teach it, so I guess we don't know it. The fact is, space is the one source of all energy in the Universe because there is only one space in the Universe. Is this taught anywhere? No, it is not. I just Google'd energy and here's what Wikipedia says: "In physics, energy is an indirectly observed quantity that is often under—stood as the ability of a physical system to do work on other physical systems. Since work is defined as a force acting through a distance (a length of space), energy is always equivalent to the ability to exert pulls or pushes against the basic forces of nature, along a path of a certain length." ((Oh look! . . . space is used to define distance . . . it is nothing but a measurement!!))

From this popular definition we can glean that science knows what energy does but it does not know what energy is. And that's what the problem behind global warming really is: lack of understanding. Apparently, popular science satisfies itself with measuring effects well enough to predict other effects and using the eloquence and thoroughness of their descriptions to replace cause with probable cause; or replace principle with theory. But theory isn't going to heal anyone of their degenerative illness, except by accident or in an indirect way. Because energy deficiency is the cause of degenerative illness, it is by increasing a person's energy that we heal them and this is the principle of allowing more intelligent energy to flow in response to sympathetic vibration, a known principle in physics.

Do you know how sympathetic vibration works? It's a simple process to understand. Did you ever see the special effect that the opera singer's voice has on the crystal goblet? She hits a note so high and with such power, that the glass shatters. The molecular structure of the crystal absorbed the vibrations of the sound of her voice and vibrated with similar frequency, but it could not handle the amplitude of her voice, so it just shattered. Another less dramatic example can be seen with tuning forks. If two are of the same frequency, you can strike one and hold it close to the other and watch it start to vibrate without having to make contact. That is the vibration in one object that is caused by a similar vibration in another object.

Now, as items of curiosity, I have two questions to ponder. What right does one person have to try to heal another? Certainly autonomy—the right to be who or what we choose to be—is the most fundamental of all human rights. So how can one person presume to invade the independence of another person's life by offering to heal them? That's question number one and number two is like it, except that it asks, what right does one person have to cohere the random background fluctuation of the energy of the space that contains another person? In other words, why would space respond in a positive way to the intentions of one person for another?

To answer the first question, all you have to do is be very much in love with another person and just see how you feel when they get hurt. It's an overpowering emotion. All you want to do is help. Help ease the pain. Call a doctor. Do something to help because when you care for someone outside yourself, you are linked with them emotionally and, therefore, vibrationally. You are excited to help the loved one because you want to ease their pain and also because you know that you can help in some way. Just being there to show you care has great value for the loved one that is suffering. It has great value for the one who is suffering, whether they are a "loved one" or not. ((Or are we all "loved ones?"))

Question number two is answered by reminding ourselves of what space truly is. First of all it is everywhere. And it is in everything. It is the way we humans perceive energy. And there is only one space in the Universe. So it has created everything. That means, if you and I are emotional, so is space. If we feel the need to help someone who is suffering, it must mean that space itself feels the need to help, too. It wants its random background fluctuation to be cohered so it honors any and all attempts to do so, delivering back the healing we seek in direct response to the power with which we deliver the healing vibration from within our own conviction.

I am glad space is vibrationally eager to mend its own creations because I wouldn't want to have to do it all on my own. It seems as though there is a symphony of vibrations that are common to all life. It is the vibration of Mother Earth and all the life she sustains and the power of life flows in the general direction of all good and all progress. We are an example of the way the Universe works. We are whole and happy until there is some incident which damages us in some way. We get knocked down, but we get back up again. We learn. We heal. We progress. There is change. But the wholeness, the harmony, is never changed and it is like the Royal Choir Of Life, still singing, waiting for our voice to chime back in again.

Everything is energy. All accumulations of matter are energy, trapped by light, into form. $E=mc^2$. The energy of your body is equal to your mass times the speed of light squared. That is a vast amount of energy locked up inside the molecular structure of your form. And everything in the Universe is either energy or it is energy, trapped by light, into form. All this energy and matter is floating around in the one intelligent space that fills this Universe. And everything is vibrating. When you and I meet, you feel some of my vibes and I feel some of your vibes. The Law of Infinite Exchange says that all accumulations of matter exchange vibrations with all the other accumulations of matter that surround them and with the background energy of space that contains them. Thus, in the creative process we are harmonizing with the universal life force that pervades all space and with that which we create. The healing process is the return to normal. We are sick because some portion of our vibration has been interfered with. An example is any degenerative illness. For some reason an energy deficiency has evolved in our otherwise perfect system of energy exchange with the surrounding Universe. By re-harmonizing with the energy of the space that contains us, and is contained within us, we overpower the deficiency and we get well.

Since it is the harmony of the Universe that heals us through the constant exchange of intelligent energy with the space that surrounds us, that implies that some of the healing thoughts are our own and some of the healing thoughts belong to the Universe which contains us—and which is thinking through us. So, my friend, when we are out to get well, if we tune our receptors in correctly, the energy of the stars will be with us to help us heal.

Eighth Chapter

DISPROOF OF THE BIG BANG THEORY

The goal of this book is to present the cosmology that enables the entire human population to participate in raising our awareness to the level where we can function as a Type 1 Planet; i.e., where we can control the energy of the Planet Earth. Once we accomplish that, we are very close to being able to control the energy of our Solar System, since all the planets obey the same rules that govern the forces that work on and through our Mother Earth. What does it mean to control the energy of our Solar System? It means we will colonize the planets and their moons and we will travel about from orbit to orbit at velocities which today, we cannot reasonably plan on because we do not currently have the theoretical physics in place that accommodates using the power of space to move through space. Right now the most popular cosmology doesn't even know what space is, yet it presumes to tell us all about the origin of the Universe. Right now the most popular cosmology is on the outside of space looking in, as it were; attempting to invent the theoretical basis for our existence in and among the stars. Where we need to be is on the inside of our understanding of space, looking out. Then what does it mean to control the energy of our planet? It means being able to control the forces of nature. These forces all act out their consequences to our environment within the space which contains that environment. Hence, knowing how to control space is how we learn to control the forces acting through space.

What is the philosophy of life we are all born into? For that matter, what is a philosophy of life? Again, from Wikipedia, we have

Philosophy of Life: There are at least two senses in which the term philosophy is used, a formal and an informal sense. In the formal sense philosophy is an academic study of the fields metaphysics, ethics, epistemology, logic and social philosophy. One's **"philosophy of life"** is a philosophy in the informal sense, as a way of life whose focus is resolving the existential questions about the human condition.

We are all participating in a way of life that is functioning when we are born into this human existence. I was born in a small town in the Catskill Mountains of New York, just two hours away from New York City. I grew up on a lake that was popular during the summer and I had friends from The City. I was from the country. Growing up I was a little envious of the city kids because they had newer cars, nicer clothes and more money. Now decades later those things still matter, but not as much as the advantage of growing up surrounded by trees instead of being surrounded by skyscrapers. At least that is my perspective today. I have enjoyed the rural way of life and the urban way of life. For me the former is better for seclusion and the resolve needed to answer the existential questions, while the latter is better for financing and embellishing the human condition. But the point is, wherever we are born, Mom and Dad are living the lifestyle we are born into. That lifestyle is different from town to town and from one society to another. Let us observe that as different as it is to live in a palace in England or in a tribe somewhere along the Amazon River, both of these lifestyles resemble each other quite closely in terms of the satisfaction of human needs. We all need to breathe, no matter where we are born. We all need to eat and drink. We have to sleep, whether on a fancy mattress or a soft patch of ground moss.

After the physical needs are met, we have emotional and psychological demands we strive to satisfy. We all need to be loved and we need to love someone else besides ourselves. We need to procreate and contribute to our family and to our society, which is part of the human family. The philosophy of life all individuals and all societies adhere to is the one that enables the species to continue to evolve on the planet. Add to that the universal language of money, which enables commerce and coexistence, and we have a good chance to make something really significant out of this beautiful blue orb of life that orbits through space. What I'd like to know is what we all have in common on that most fundamental survival level so the prescribed metaphysical remedy for the planet's problems will transcend social barriers.

We have considered the cause and effect relationship that creates our physical reality and for this argument we can think of the physical world as the demonstration of the action of the causal world. On the level of demonstration, all animals act the same way. Human beings are animals. What all animals have in common is the stimulus/response mechanism which shapes our behavior through our senses. The only belief system that works on the actual physical level of demonstration—and it is the essential belief common to all animals—is the presumption that life is the way it appears to be.

We need to know that when we are out walking through the woods and we hear a rustling of the leaves and the sound of a rattle, that precise area where we perceived the threat of a poisonous snake needs to be avoided and by a pretty wide margin, too. Rattlesnakes are very dangerous. If it were not so, I would tell you. But what is the point? It is this. All animals perceive cause and effect on one level; i.e., the physical level of effect. The system needs to work this way because animals have no imagination. They can only see what is right in front of them. A gazelle, for example, sees a lion and all she can do is try to outrun him. She may have some sophisticated escape techniques that are stored as instincts in the gazelle's race mind, which her brain is connected to, but devising a more elaborate plan is out of the question. The only race mind that has a link to the imagination of the Universe on a causal level is that of the human race. We are special. We actually have the capacity to manipulate the effect by changing the cause. The system of the Universe is already set up like this. It is the way the Universe works, whether we know it or not; whether we use it to our own benefit or not. The problem we are having is this: because of the way we were raised, not just some of us, but all of us; because we learned according to the principles we were all born into; we are forced to presume that the most basic premise of any and all ways of life must be the most apparent of all observations: which is that life is the way it appears to be. We need to actively support and submit to that premise in order to survive. In fact, the more accurate our perception is, the better are our chances for survival. And what do our senses tell us? What is the most basic assumption hidden in this "philosophy of the obvious" that we all bow down to? Since our world is demonstrated in the three-dimensional franework of our senses, it must be that matter is solid, since it is the solidity of matter which provides the objects we measure in those three dimensions; and its attendant, the apparent emptiness of space.

These are the nuts and bolts we used to build our world, just like the birds build their nests and the lions fabricate their dens. The only problem is, neither assumption is correct. Matter is not solid and space is not empty. But this only matters to creatures that have the capacity for abstract thinking. And who has that capacity? We all do! Now that is the expression of hope: that we all have the capacity for change through the use of the imagination. At the same time, it does not seem to stand a chance because it proposes to modify our survival behavior and that is the one place where we think, "if it ain't broke, don't fix it!" Fortunately, the crack in the hull of that argument is the fact that IT IS BROKE! Watch . . . as we climb the geometric slope of the environmental change we are living in right now. It will become more and more obvious that we have to change something, somehow, because to continue the behavior that caused the problem never solves the problem. But the behavior still has to change. And this time we cannot be satisfied with just changing the effect on the effect level because that is not guaranteed to last. It has to be changed on the causal level.

On the causal level of what generates behavior, it is observed that there is some image that we have of ourselves that we are always living up to. It is a picture—a canvas—upon which the colors of love and hate, joy and sorrow, indeed all of life's experiences; are brushed by those who raise us. We are like those we grew up with for they, too, have added a brush stroke here and there. But the day comes when we must be held accountable for our own thoughts and their subsequent actions; when it must be admitted that we are the artist who sings at the bottom of the canvas every day.

It is said, the child is father of the man (or woman). In our subconscious mind we have the picture we started to paint at birth, perhaps even earlier. It is the picture of who we are today. If you ever want to know what is going on in your subconscious mind, just look at your world right now, because it is that subliminal self which creates our physical reality according to the self-image we formulate. Were we happy and loved as kids? Then we are probably pretty successful in our adult life. Is your Dad an attorney? Then you are probably an attorney. Do you have a loving Mom? Then you probably love your children. Is your Mom or Dad an alcoholic? Then there's a good chance you are, too. We cannot see the picture we are drawing of our self because we cannot see emotion or logic. It is thought which creates form and all thought consists of emotion and logic. These components form the polarity of thought. While the reasonableness of an idea is what sustains its growth it is the feeling associated with the concept that powers it. Every thought has these two components.

We also partake of the collective self-image. There are colors on our canvas that were splashed on by the kids we grew up with and we splashed some of our own colors on their canvases in return. Maybe we had a school pride growing up, maybe we didn't. It seems to me there were some neighbors we sort of idolized because of the authority of their success, and many of us had heroes we adopted from stage and screen. I remember seriously considering being a spy right after the first James Bond movie came out. I mentioned this to the recruiter I talked to and he admitted, whenever a war movie was successful, he got a flood of enlistees in the weeks that followed. We are living up to some collective image, too. And that brings me back to the Big Bang Theory.

Scientists make up only a small percentage of the population. Why is it then that in a poll done by one of the major networks, for the New Year's Eve show celebrating the beginning of the 21st century, found that the most important person of the 20th century was Albert Einstein? Clearly he is one of humanity's greatest heroes. He is someone we all admire and many of us would like to be like Albert Einstein. We admire his attributes; his wisdom;

the global influence of his thinking; his Theory of Relativity. And of course, the most famous equation of all time, E=mc^2. This is Einstein: one of the gods of physics, who can do no wrong; judged by the world to be the most important person of the 20th century. And this is the guy who is very much responsible for the birth and continued popularity of the Big Bang Theory. And we worship him as one of our greatest heroes. Because of him, the bombs were built that leveled two cities in Japan and basically stopped the war we had with that country. Because we had him and Hitler didn't, we saved the world. Okay, in a sense, he saved the world. But that's what heroes do.

What heroes also do is influence the thinking of those who worship them. Since Einstein was voted by the world to be the most important guy in the world when all these scientific discoveries were being made, you could say that our thinking is influenced quite a bit by the image of Albert Einstein that rests in the collective sub-consciousness of humanity. I'm going to risk saying that the collective consciousness of the planet believes in the validity of whatever Albert Einstein is said to have believed in. And, since his General Relativity demands that the Universe began with a big bang, then there must have been a big bang which started this Universe.

But I disagree! And so does any rational thinker who is willing to take a clear, unbiased look. The first reason we have to doubt the validity of the Big Bang Theory is that it is entirely unnecessary. You know as well as I do that there is always more than one explanation for any observed phenomenon. For example, that's what magic is, with its disproof of the validity of our senses. We saw the shell go under that particular cup and we are surprised when it is not there. We thought we saw one thing happen and we expected a particular outcome. But we were tricked. What really happened was something else. In this case, just because the Universe is expanding it does not have to follow that it all started out in one small place called a "singularity" and started to expand after a big explosion. There is another explanation which it is much easier to believe, namely that the Universe has always existed so it doesn't even have a beginning. The reason the Universe is expanding is because that's just the way it works. There is a constant influx of energy through the fabric of space from the other side of the fabric, where there is all light. Over here on this side of space, there is all creation. All material objects are particles trapped by light into form. The light is constantly coming from the other side of space into this side where we live, creating things, so the Universe has to expand to accommodate this constant influx of energy.

Isn't that a lot easier than having to try to shrink billions of galaxies down into the size of a pea? What are you Big Bangers thinking?

The second objection is the fact that the rate of expansion is accelerating. If it all started with an explosion, it should be slowing down, not speeding up. That right there should be enough to throw the Big Bang theory right out the window! But no, hero worship often blinds those who choose to be led. To explain this phenomenon and still cling to the idea of the Big Bang, physicists have invented "dark energy," a mysterious force that somehow overcomes gravity. I'm wondering, have you guys been watching too many grade B science fiction movies? The fact is, the acceleration of the rate of expansion is a natural consequence of the way in which the energy is passed from the other side of the fabric of space to this side. More specifically, because the seam is space is in the form of an arc, and particles always accelerate when they move through a curve, the rate of expansion is constantly accelerating. That should be enough contradictory information to create the doubt needed to look elsewhere, but there's more.

The third objection we have to the theory of the Big Bang comes to us in the form of Occum's Razor; otherwise known as the law of parsimony, economy or succinctness. It is a principle stating that among competing hypotheses, the one which makes the fewest assumptions should be accepted. Again, this definition comes from Wikipedia, which goes on to say that in the scientific method, succinctness is not considered an irrefutable principle of logic, and we are not considering it as such. Occum's Razor is used as a heuristic, which is a technique of problem solving that uses such humble mechanisms as an educated guess or good old common sense.

Let's digress for a moment. What is common sense? Is that where we swallow somebody else's argument hook, line and sinker just because they assure us that they are smarter than we are and they have the degrees to prove it? Even though the concept sounds contrived? When I first heard the theory in seventh grade science class, I knew there was something suspicious going on. That was fifty years ago. The Universe has gotten a lot bigger in fifty years. But physicists still expect us to believe that the Universe started out as small as a tiny pea-sized something-or-other and then one day . . . KABOOM!! . . . WOW!! . . . A BIG EXPLOSION!! And then the little teeny tiny pea didn't just become a giant star or even a vast galaxy. Oh no . . . that's nothing, Dear Sir or Madam who is reading this. That teeny tiny pea ended up becoming billions of galaxies, each with billions of stars. In fact, we don't even know how many stars. Well, actually we do know exactly how many stars there are. The number is infinite. So, we do not know how big the Universe is because we haven't found the edge of the Universe and Guess what? We never will. We just keep finding more and more Universe because it is infinite. So how can you get an infinite quantity from a finite beginning? You can't. The Big Bang theory is a farce. But all the smart scientists out there think we're stupid enough to believe whatever they hand us.

And so far, they are correct. But here is objection number five. And this is the big one; the decisive blow; the straw that broke the camel's back. Their misconception is killing us. Oh, it's killing them, too. As far as I know, degenerative illness strikes as often in the scientific community as in any other.

Minds that cling to the finite Universe have not been able to clear themselves of the effects of the misconceived perception of the solidity of matter and the emptiness of space. To those who believe wholeheartedly in the validity of what they perceive is given the responsibility of making sure everybody else agrees with them because that's the only reward you are ever going to get out of believing in something that is not true. They have to make a noise louder than anybody else because they know they are wrong and they continue to believe in a lie. They need the added energy of anxiety that comes with all neurotic behavior.

This final spear of truth we thrust into the Big Bang Beast is the most important because it is the difference between life and death; life for the person suffering from the torment of degenerative illness through death of the self-inflicted lazy way of thinking what everyone else indulges. It's okay to keep Einstein as a hero but it is bad for our health to believe in a finite Universe. First, because it just is not true and second, because it is the one subtle truth that influences the flow of healing energy to us and through us.

We're way out there now, fellow traveler; out in the farthest reaches of space, which we found by traveling into the infinite reaches of space within. We are infinite in both directions because space is infinite. We are introducing the New Science of Space as a remedy for all degenerative illness because this is the time now in the evolution of our species here on Earth to take the quantum leap from living as mere physical beings in a finite world of matter, to living as centers of energy in a Universe of infinite intelligence. If you understand the difference then you can also visualize the possibilities that await us as developers of the Solar System, which we become by learning to control the energy of our planet Earth.

Information has been provided to substantiate the following closing argument. The New Science of Space is based upon the cosmology that is being presented as a replacement for the Big Bang Theory.

This is our first contention.

Energy deficiency is the root cause of all degenerative illness.

This is our second contention.

The energy level of the human body is mediated by the space that contains us and is contained within us.

This is our third contention.

Any affiliation with the Big Bang Theory implies belief in a finite Universe, which automatically shuts off the flow of a portion of the intelligent energy that is exchanged between our body and the space which contains us and is contained within us.

This is our fourth contention.

The lack of intelligent energy flow is the deficiency which leaves the human body unprotected from the random occurrence of degenerative illness.

Therefore, we are driven to our fifth and final contention.

By changing the individual's energy system from finite to infinite potential, we avail the body of the sympathetic vibration needed to overpower the illness; i.e., to be healed. In other words, we use information to cure degenerative illness.

EPILOGUE

Fellow Travelers . . .

The Spirit of the Planet is speeding down the wrong set of tracks like a runaway freight train. Because of the *learned* error in our thinking, we continue to fuel that motion with our belief in the validity of our own perception, even though we do not like what we see happening to our planet. We have not been taught that what we see is our own concept. We think the world . . . or the planet . . . or the environment . . . happens to us. The truth is, we are responsible for everything that happens within the space that contains our planet and everything on it. We learn to control the energy of the planet by first learning to control ourselves.

We are all space cadets; trainees in learning the art and science of cohering the random background fluctuation of the energy of the space that contains each of us individually and then collectively; since the paths of our vibrations are constantly entangled with everyone else's. The space that contains you is waiting patiently to hear your first "Hello." Then at some point further down the path you are traveling, probably when you least expect, you will sense its "Hello" in return; in the form of a healing, or an inspiration, or an event of unparalleled importance in your life. It may be the presence of all light in a flash, light lightning in your soul, but whatever it is, you will know. And you will be glad.

Eric A. Staubes

Ph.D. in Metaphysics; Metaphysical Practitioner;

Fellow of the Holistic Life Coaches Association; Space Coach

The discovery about SPACE came to me in New York in 1979, like a Virgin Birth from the Cosmos; and I kept it secret for 33 years while it developed into a Force for Good. WHAT I LEARNED THEY DO NOT TEACH IN ANY SCHOOL, so I had to go it alone. It is our duty, TEAM EARTH, to control the energy of this planet. I believe that once we all know the truth about SPACE, we will join forces and know what to do next.

At Your Service

Coach Staubes

www.ingramcontent.com/pod-product-compliance
Lightning Source LLC
Chambersburg PA
CBHW021928170526
45157CB00005B/2239